ONE SMALL SQUARE®

The Night Sky

by Donald M. Silver

illustrated by Patricia J. Wynne

LEARNING
TRIANGLE
PRESS

*Connecting kids, parents, and teachers
through learning*

An imprint of McGraw-Hill

New York San Francisco Washington, D.C. Auckland Bogotá
Caracas Lisbon London Madrid Mexico City Milan
Montreal New Delhi San Juan Singapore
Sydney Tokyo Toronto

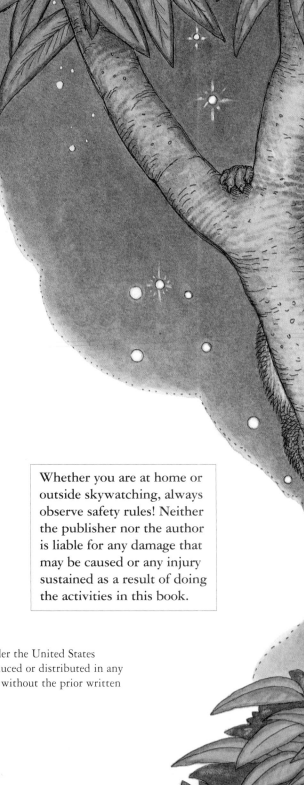

If you come to a word you don't know or can't pronounce, look for it on pages 44-47. Answers for the activities will vary.

For my mother Jewell Wood Wynne

—my first guide to the wonders of starry nights

We are very grateful to Dr. Ed Krupp of the Griffith Observatory in Los Angeles for his thoughts and comments on the night sky. We also extend our sincere appreciation (yet again) to Marc Gave, Ivy Sky Rutzky, Maceo Mitchell, and Thomas L. Cathey for their contributions to this star-studded addition to the One Small Square series.

Whether you are at home or outside skywatching, always observe safety rules! Neither the publisher nor the author is liable for any damage that may be caused or any injury sustained as a result of doing the activities in this book.

Text copyright © 1998 Donald M. Silver.
Illustrations copyright © 1998 Patricia J. Wynne.
All rights reserved.
One Small Square® is the registered trademark of Donald M. Silver and Patricia J. Wynne.

Library of Congress Cataloging Number 97-075976
ISBN 0-07-058045-6
    4 5 6 7 8 9 0 QPD/QPD 9 0 3 2 1 0

Introduction

Come explore the night sky. It's your window on the universe. Look long. Look deep. What you see will boggle your mind!

What's out there? Blazing heat and brutal cold. Blinding light and total darkness. Exploding stars and vast emptiness. And that's just for starters.

Gaze at Mars, where robot spacecraft have touched down for a close-up look at the Red Planet. Watch space rocks streak across the sky, only to burn up before your very eyes. Stare at light that left stars before Columbus discovered America. Catch a glimpse of a great comet hurling through space toward the Sun. Better yet, be the first to sight a new comet.

Polaris

Lost? On a clear night, you can find north once you recognize the stars that make up the Big Dipper. It looks like a cup with a handle. At the front end of the cup are two stars. Imagine a line connecting them. Extend this line five times and you'll be at Polaris, the North Star. Face Polaris and you face north. Any way you see the Big Dipper—right side up, upside down, or on its side—this line always points to Polaris.

Polaris

Pointer stars

You can explore the night sky from just about anywhere you can see it—the city, the country, a mountaintop, a ship at sea, or your own backyard. You don't need binoculars, but if you own a pair, bring them along. They will help you zero in on the Moon's mountains and craters.

To probe the planets or spy deep into space, you'll need a telescope. This book will explain how you might locate one you can use.

No matter how or when you view the sky, there is one rule you must not forget: NEVER EVER look directly at the Sun. Its blinding light can damage your eyes.

Also, don't forget that once every 24 hours, the Earth turns completely around from west to east. So by day, the Sun seems to rise in the east, move across the sky, and set in the west. At night, most of the stars seem to do the same.

The night sky is way too big to explore all at once. First you need to know how to tell stars apart. Then, where to look for planets and moons. And then, what a nebula is. You'll find these and lots more in one small square. Soon you'll know your way around the night sky. One day you may bring up on your computer a message from astronomers, asking you to skywatch and report back over the Internet what you saw.

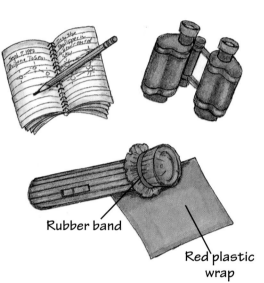

On a clear night, you'll be able to see plenty of other stars with just your eyes. But with binoculars or a telescope, many thousands more will appear. Carry a flashlight with red plastic wrap over the front end. Red light won't lessen your night vision but will let you read a sky chart.

Rubber band

Red plastic wrap

As the Earth turns, Polaris, the North Star, is the only star in the night sky that stays in one place. Groups of stars, such as the Big Dipper, seem to circle around it. Still others seem to rise, move across the sky, and set before morning.

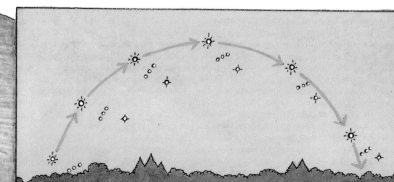

1. ALWAYS get permission to go out at night. ALWAYS tell someone where you are going and when you expect to return.

2. ALWAYS take along a jacket or a sweater. Nights turn chilly, even in summer. In winter, dress really warmly, with gloves and a hat.

3. ALWAYS use your flashlight and watch your step in the dark.

4. Use houses, trees, fences, or poles below your viewer to help you locate your small square each night.

5. Give your eyes 15–20 minutes to get used to the dark. During this time don't look at any bright light. If you do, your eyes will have to readjust. When you need to read a sky chart or take notes, use your red-tinted flashlight.

6. You may find it easier to stargaze while lying on an old sheet or blanket or on a lawn chair.

One Small Square of Night Sky

After sunset, if you're allowed, go out and look at the darkening sky. You can choose any part of the night sky to explore. The sky charts on pages 40–43 will help you figure out which stars are in your small square.

The square that this book explores is shown on the facing page. Around it is a ruler viewer you can copy on cardboard or heavy paper. Cut out the center so that when you hold the viewer up, you can see one small square of night sky inside it.

Wherever you live, this small square is one of the easiest to find from January to April. That's because it contains some of the brightest stars in the night sky. Not even the glare of lights in a big city can hide these stars on a clear night.

Before you go out to stargaze, read Skywatch on this page. Once outside, try to locate this square. In the northern hemisphere, look south. In the southern hemisphere, look north. In both cases, search for the brightest star in the sky. Hold your viewer so that this star is between 1 and 2 on the bottom ruler, as shown. Then move your eyes up from the star and search for three bright stars in a row. Finally, turn your viewer to include the other stars in the picture. To make your square smaller, hold the viewer at arm's length.

Follow along as One Small Square unlocks some of the secrets of the Bull, the Belt, and "Beetlejuice." When evening comes, don't reach for the TV remote. Don't reach for your mouse. Reach for the stars.

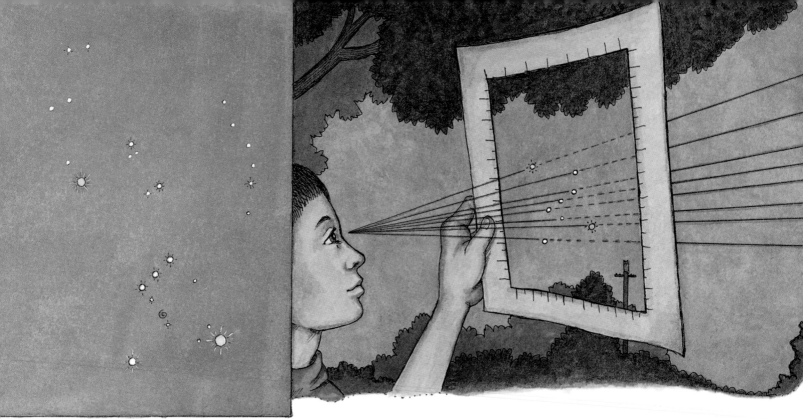

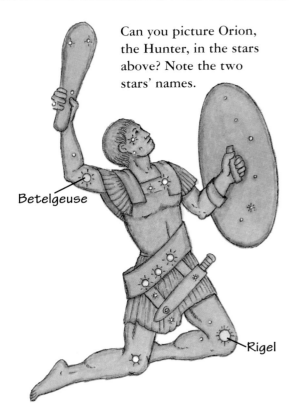

Can you picture Orion, the Hunter, in the stars above? Note the two stars' names.

Betelgeuse

Rigel

Seeing Stars

Look! Up in the night sky! It's a bird. It's a plane. It's anything you can picture. You can connect stars like dots in a game book and create fun shapes and patterns.

Take the three bright stars all in a row near the center of the small square. Can you imagine them as three mountain sheep? Some American Indians could. Or that the three stars are lined up on a turtle's shell? That's what Mayan Indians in Mexico saw. To some native Australians, the stars were three fishermen in a canoe. To the ancient Greeks, they marked Orion's belt.

Who was Orion? In Greek myths he was a mighty hunter. He angered Mother Earth by boasting that he could kill any of her beasts, large or small. So she sent a monster-sized scorpion to sting Orion to death. The mighty hunter attacked but could not escape the scorpi-

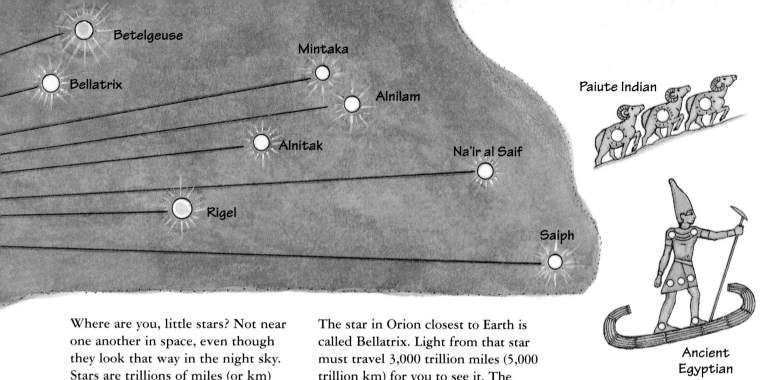

Betelgeuse

Bellatrix

Mintaka

Alnilam

Alnitak

Na'ir al Saif

Rigel

Saiph

Paiute Indian

Ancient Egyptian

Hindu

Old German

Maya

Where are you, little stars? Not near one another in space, even though they look that way in the night sky. Stars are trillions of miles (or km) away from Earth—and usually from each other. Only the Sun is close. That is, if you call 93 million miles (150 million km) close!

The star in Orion closest to Earth is called Bellatrix. Light from that star must travel 3,000 trillion miles (5,000 trillion km) for you to see it. The journey takes about 500 years. The Bellatrix you see today is the way it was 500 years ago.

on's poison. To honor Orion, the gods placed his image in the night sky. They set the scorpion there, too, but far enough away that the deadly stinger could never again reach the hunter.

Today, nobody believes the story of Orion. Nor that the night sky is really home to hunters and monsters. But the stars that the ancient Greeks saw as Orion are still there. They make up one of 88 star groups called constellations into which the entire night sky can be divided. All the stars in the constellation Orion are in the small square. Parts of 6 other constellations are there too. They are the Twins, the Unicorn, the Big Dog, the Hare, the River, and the Bull. Imagine them in your mind before you turn the page and have a look.

Could the three stars in Orion's belt be sheep? Part of the Egyptian god Osiris? An arrow? Three farmers? A turtle shell? Why not all? That's what they have been to different peoples at different times. What do you see?

Pollux

Castor

The Twins
(Gemini)

In this book the brightest stars in the night sky are drawn the biggest. Dimmer stars are shown smaller. Constellations with at least one bright star are easiest to find.

The Hare (Lepus)

The Unicorn
(Monoceros)

The Big Dog
(Canis Major)

Sirius

Adhara

Small square

Use these clues to find constellations: Orion, beware. The Bull's ahead, the Unicorn behind. The Twins above your club are scared. Call your Big Dog from hunting the Hare. And don't fall in the River.

The best way to pick out constellations in the night sky is with your eyes alone. If you try binoculars or a telescope, you'll be able to look at only a few stars at a time instead of the entire star picture.

Carry this book with you when you skywatch. Try to match the star patterns pictured in it with the stars in the sky. Remember: You'll need your red-tinted flashlight to read in the dark. Without good night vision you'll miss the dimmer stars that help form nearly every constellation. If you like, expose glow-in-the-dark stars to light and stick them on top of the patterns in this book. In the dark, when you open to a pattern, it will glow for an hour or so as you search the sky for it.

Once you get to know the pattern of stars in a constellation, you can find it much more easily. It

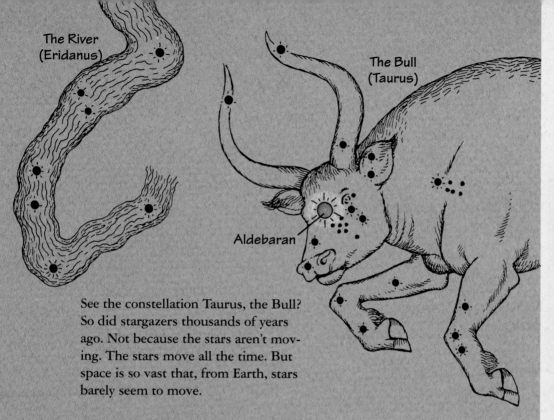

The River
(Eridanus)

The Bull
(Taurus)

Aldebaran

See the constellation Taurus, the Bull?
So did stargazers thousands of years
ago. Not because the stars aren't moving. The stars move all the time. But
space is so vast that, from Earth, stars
barely seem to move.

can help you locate other nearby constellations. For instance,
the three stars in Orion's belt point in one direction to the
Bull, and in the other direction to the Big Dog.

Sometimes a single star does the trick. Sirius, the Dog Star,
shines brighter than any other in the night sky. If you find it,
you'll find the Big Dog. Orion is just a star hop away.

If Orion is in your small square, you may see it sideways,
upside down, or right side up. It all depends on where you are,
where your square is in the night sky, and what time you are
out stargazing. As the Earth turns, Orion may rise in your
square or move into and then out of it. With your viewer, you
can follow him across the sky. Or you can keep your viewer in
one place and see how your square changes from one hour to
the next. No matter what you decide, take time and get to
know the stars in your small square. Zoom in for close-ups
with binoculars. Really look—you'll see stars.

Your Night Sky Notebook

Exploring the night sky is like
exploring another world. It's
full of stars, planets, moons,
comets, and endless other
wonders of the universe.
Who knows what you may
find in your small square on
any given night? That's why
you always need to carry a
notebook and a pen or a pencil
with you when you skywatch.

Once you choose your square,
draw a diagram of it in your
notebook. Record the date,
the time, and the place, as
well as how clear or hazy
the sky is. Which stars are
brightest? Which are hardest
to see? What did you use to
watch the sky? Just your eyes?
Binoculars? A telescope?

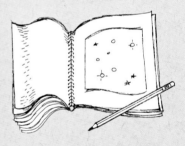

Connect the stars in your
diagram to create your own
constellations in addition to
those on pages 40–43. Name
your constellations as you like.
Does your square change during the time you are outside?
If so, how? Go out on another
night at exactly the same time
and redraw your square. Does
it match your first drawing?
Can you figure out how your
square changes every night?

What's Your Angle?

How far apart are the stars in the night sky? Scientists find out by using angles. Angles are measured in degrees. You can use angles too. Just stretch out your arm, close one eye, and raise your index finger. If that finger fits between two stars, they are about 1 degree apart. If your fist separates the stars, they are about 10 degrees apart. If you spread out the fingers on your hand, there are about 20 degrees from the tip of your thumb to the tip of your little finger. Now, go to work. How far apart are Betelgeuse and Rigel? Rigel and Sirius? The stars in your square? Use both hands if you need to and record the angles in your notebook.

Count Your Stars

Compared to millions of other stars, the Sun isn't all that big or bright. Indeed, many stars in the night sky are just like it. But being so distant, their light is too dim to see. Or is it? Count the stars you see in your square. Recount them using binoculars. How many do you think you would find with a telescope?

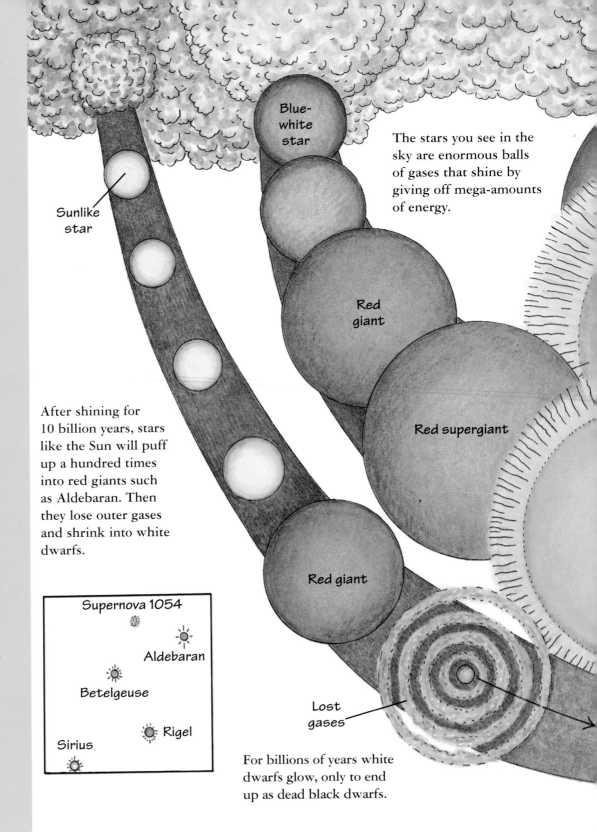

Sunlike star

Blue-white star

The stars you see in the sky are enormous balls of gases that shine by giving off mega-amounts of energy.

Red giant

Red supergiant

After shining for 10 billion years, stars like the Sun will puff up a hundred times into red giants such as Aldebaran. Then they lose outer gases and shrink into white dwarfs.

Red giant

Lost gases

For billions of years white dwarfs glow, only to end up as dead black dwarfs.

Supernova 1054

Aldebaran

Betelgeuse

Rigel

Sirius

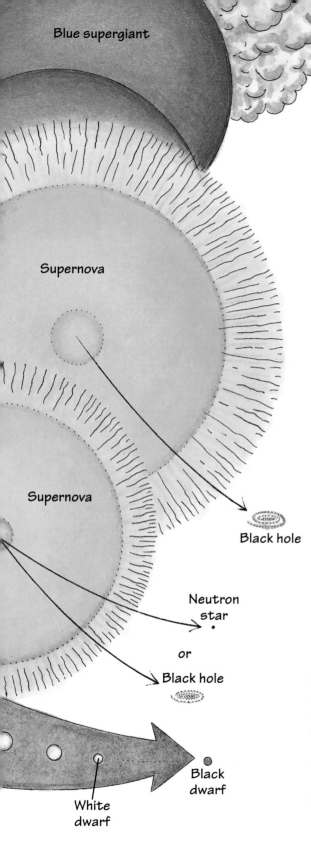

Blue supergiant

Supernova

Supernova

Black hole

Neutron star

.

or

Black hole

Black dwarf

White dwarf

Starlight, Star Bright

Do you ever make a wish upon the first star you see at night? That star may be Rigel, in Orion's left knee. Or Betelgeuse, in his raised right arm (the star many people pronounce as "Beetlejuice"). Whatever the star, the next time you make your wish, take a long look at it. Do the same for all the stars in your small square. Seeing is believing: Stars shine in different colors.

Watch Rigel sparkle electric blue-white. It's a much bigger and hotter star than our yellow-white Sun. In fact, 50,000 of our Sun joined together would not shine as brightly as Rigel.

The reddish glow of Betelgeuse is a clue that it is cooler than the Sun. But it is huge! If it took the Sun's place, it would stretch past Earth to Mars!

How can a supergiant like Betelgeuse be just a point of light in the night sky? And how can it be safe to gaze at Rigel's light but not the Sun's? Because Betelgeuse and Rigel are so very, very far from Earth, they appear tiny. Their light can't hurt you. You also can't tell how bright they really are. If only you could wish them next to the Dog Star, you'd know in an instant. Hot, big blue Rigel beats cool, big red Betelgeuse, and both far out-shine the brightest star in the night sky. Have you ever made a wish on it?

Nothing lasts forever, not even stars. Those such as Rigel shine "only" millions of years as very hot, blue-white stars. Then they grow into red supergiants such as Betelgeuse. But these red stars are doomed to collapse and blow apart in the biggest of all explosions—a supernova. If you had lived in the year 1054, you could have seen a supernova in the small square. That blast left behind a tiny neutron star maybe 10 miles (16 km) wide. Other supernovas leave mysterious black holes from which nothing, not even light, can escape.

Block That Light

How can one star block its partner's light? That depends on how big and bright each star is. Take two flashlights and cover the end of one with red cellophane. Give each flashlight to a friend. Ask your friends to circle each other, holding the flashlights so the beams shine away from them.

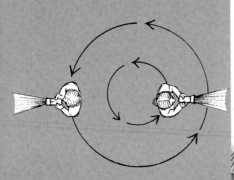

Stand far enough away so you can see each beam clearly when it shines in your direction like starlight. What happens when one friend winds up directly in front of the other? Draw what you see in your notebook. What happens when they switch positions? Have one friend stand and the other crawl as they circle. Does that change what you see?

On the Rise

Even though you know that stars really don't twinkle, try this: Early one evening, watch stars rise in the sky. Where do they appear to twinkle the most—nearer the ground or higher up? Where the least? Can you figure out why? (Hint: Higher up the air is thinner.)

In a Twinkle

Twinkle, twinkle, little star. Most likely you know the rest. Just about everyone does. And just about everyone has seen stars twinkle at night. But when flickering starlight in your small square catches your eye, don't think twinkle, twinkle. Instead, think of winds, storms, and heat rising from the ground. These cause pockets of air to shake. When starlight passes through shaky air, it appears to be twinkling.

So forget twinkle. After all, the Sun is a star and it doesn't twinkle. It shines with a steady light, as most of the stars in your square do. Every night that you explore the sky, take time to get to know the light of different stars. Only then will you be ready to double your fun.

Many of the stars in the night sky are not alone. They have partners. Each partner circles the other. At

If a dim star circles a bright star and passes in front of it for a short time, some of the bright star's light may be blocked.

Partner star

Do you see any change in the brightness of Betelgeuse from one week to the next? Betelgeuse is one of those stars that shrinks and swells. Its partner is much smaller than the supergiant.

Betelgeuse

Dim partner

Bright partner

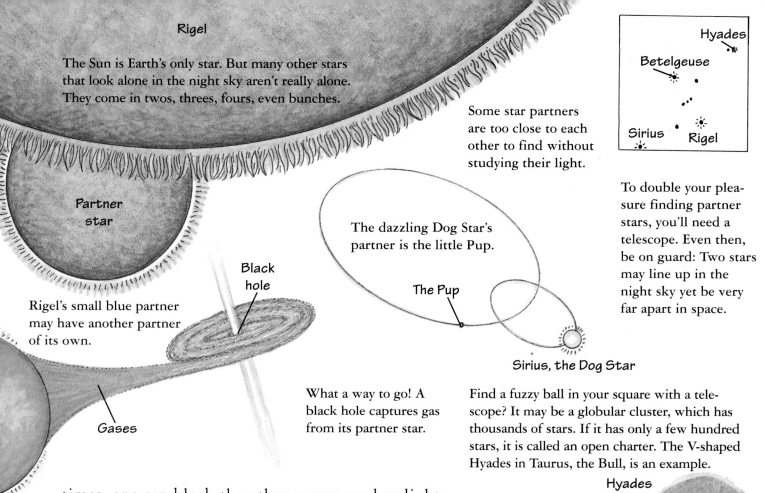

Rigel

The Sun is Earth's only star. But many other stars that look alone in the night sky aren't really alone. They come in twos, threes, fours, even bunches.

Partner star

Rigel's small blue partner may have another partner of its own.

Gases

Black hole

What a way to go! A black hole captures gas from its partner star.

Some star partners are too close to each other to find without studying their light.

The dazzling Dog Star's partner is the little Pup.

The Pup

Sirius, the Dog Star

To double your pleasure finding partner stars, you'll need a telescope. Even then, be on guard: Two stars may line up in the night sky yet be very far apart in space.

Hyades
Betelgeuse
Sirius
Rigel

Find a fuzzy ball in your square with a telescope? It may be a globular cluster, which has thousands of stars. If it has only a few hundred stars, it is called an open charter. The V-shaped Hyades in Taurus, the Bull, is an example.

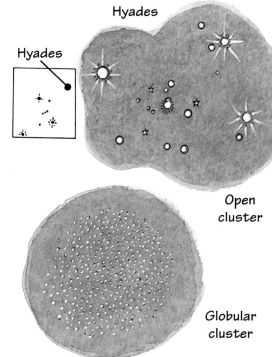

Hyades

Hyades

Open cluster

Globular cluster

times, one can block the other so you see less light coming from that spot in your square. Note where the spot is and how long it remains dimmer than usual. Then get out binoculars or a telescope and try to find both star partners.

Splitting one star into two partners is a real challenge for any skywatcher. What looks like one star may turn out to be a patch of hundreds of stars sparkling like diamonds. Or it may be only one star that shrinks and swells over days, months, or years. Such a star dims when it shrinks and brightens when it swells again. You'll never forget the moment you discover partners, patches, or other unusual stars.

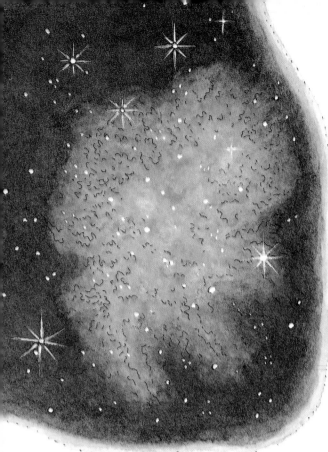

Where Stars Are Born

It's not your eyes. And it's not your binoculars, so don't bother to clean them again. There's nothing you can do to make that star below Orion's belt less fuzzy-looking.

Three bright stars hang from the belt like a sword or a dagger. The fuzzy one is in the middle. It has an air of mystery about it and well it should. It may shine like a star but it's way more than that. It's the Great Orion Nebula, a place where star birth occurs.

A nebula is a dark cloud of gas and cosmic dust between certain stars. Such a cloud may just sit in space for billions of years. Then shock waves from a supernova may cause part of the cloud to clump together, knot up, and shrink. The more it shrinks, the more the gases and dust pull in, collapse, and press each other. Under pressure they slowly heat: hot, hotter, superhot. When the temperature at the center reaches a few million degrees, a shining star is born. Tremendous light and heat energy rise to the surface and travel into space in all directions.

The Great Orion Nebula may produce 10,000 stars like the Sun. Many already have been born, both there and in the small square's other stunning nebulas. Someday you may see a nebula through a telescope. Until then, even a peek at where stars are born is awesome.

No light shines from this gas-and-dust cloud. In fact, it is so dark that it blocks the glow from more distant stars. Where light peeks over the cloud's towering edges, it's easy to see why the cloud is called the Horsehead Nebula.

Step back in time. The year is 1054. A new star appears in Taurus, the Bull. It shines so brightly that for weeks it can be seen both night and day. People living in China write about it. American Indians may have painted it on rocks. Only it wasn't a new star being born but a dying star self-destructing in a supernova. Today, with a small telescope you can see some of the star's outer layers that shot into space during that blast. They formed the Crab Nebula, which still expands millions of miles (km) every day.

Crab Nebula

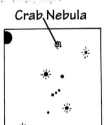

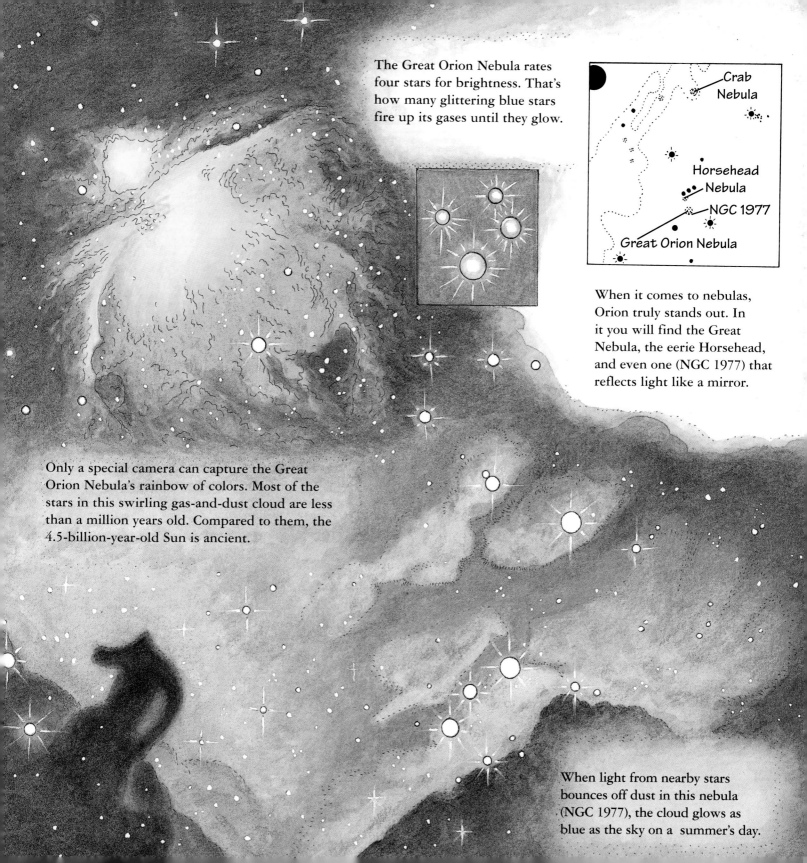

The Great Orion Nebula rates four stars for brightness. That's how many glittering blue stars fire up its gases until they glow.

Crab Nebula

Horsehead Nebula

NGC 1977

Great Orion Nebula

When it comes to nebulas, Orion truly stands out. In it you will find the Great Nebula, the eerie Horsehead, and even one (NGC 1977) that reflects light like a mirror.

Only a special camera can capture the Great Orion Nebula's rainbow of colors. Most of the stars in this swirling gas-and-dust cloud are less than a million years old. Compared to them, the 4.5-billion-year-old Sun is ancient.

When light from nearby stars bounces off dust in this nebula (NGC 1977), the cloud glows as blue as the sky on a summer's day.

Irregular galaxy

Spiral galaxy

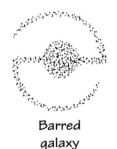

Barred galaxy

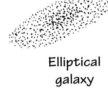

Elliptical galaxy

There are galaxies galore in the universe. Take your pick. Some have no special shape—they are irregular galaxies. Others are pinwheel-shaped spirals, like the Milky Way. Then there are barred galaxies. They have a bar of stars along their center, with spiral arms poking out the ends. Finally, there are football-shaped—elliptical—galaxies. Some of these have ten times more stars than the Milky Way!

Special as the Milky Way is to Earthlings, it is but one of billions of galaxies.

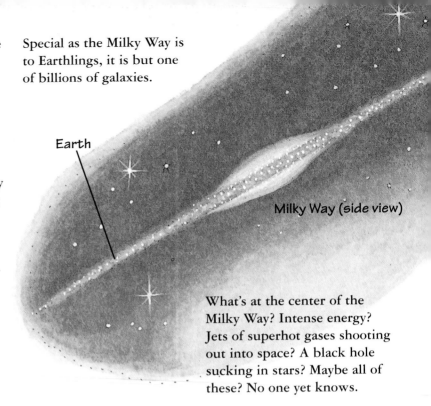

Earth

Milky Way (side view)

What's at the center of the Milky Way? Intense energy? Jets of superhot gases shooting out into space? A black hole sucking in stars? Maybe all of these? No one yet knows.

Home

What do you think of as home? Your room? Your house or apartment? Your town, city, state, country? Maybe the Earth? All of these? Whichever ones fit the bill, don't forget the Milky Way Galaxy, of which Earth is a part. It's your home in the universe.

Want to see it? You'll need to be far away from any city's bright lights. If the night is clear and moonless, scan the sky for a faint glowing band of milky-white light arching across it. That's the part of the Galaxy we can see from Earth. It passes through the small square. It may also pass through yours.

At first glance, the Milky Way may not seem a big deal. But you'll quickly change your mind when you focus in on it with binoculars or a telescope. Everywhere you look, your eyes will feast on stars by the thousands. And they are just a drop in the bucket. There may be up to 200 billion stars in

Milky Way Galaxy

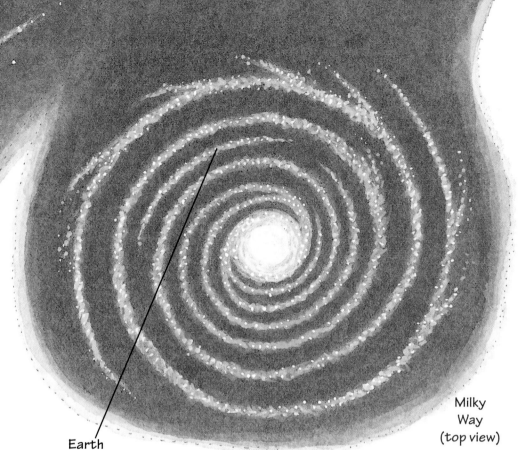

Earth

Milky Way (top view)

By far, the Hubble Space Telescope (above) commands the best view of the universe. No clouds, no smog, no city lights get in its way 370 miles (596 km) above you. Look (below) at what Hubble discovered in one corner of space: layers and layers of galaxies.

the Milky Way. Nearly everything you can see in the night sky is part of it.

Don't expect to figure out the true shape of your home galaxy by looking at it from Earth. You are seeing the Milky Way from the side, edge on. It appears to be like two fried eggs glued together back to back. Notice any dark patches? They aren't holes in the Galaxy but thick nebulas. Talk about getting in the way! These thick gas-and-dust clouds totally block your view of the Milky Way's star-filled center.

To see the entire Milky Way, you would have to sail off into deep space. From there, high above it, you would draw the Galaxy as a pinwheel with a big disk at the center. The Sun would be a dot toward the edge of one of the pinwheel's spiral arms. And Earth, a speck of dust you call home.

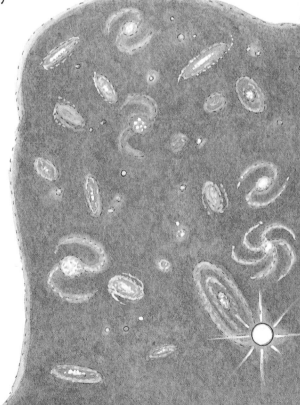

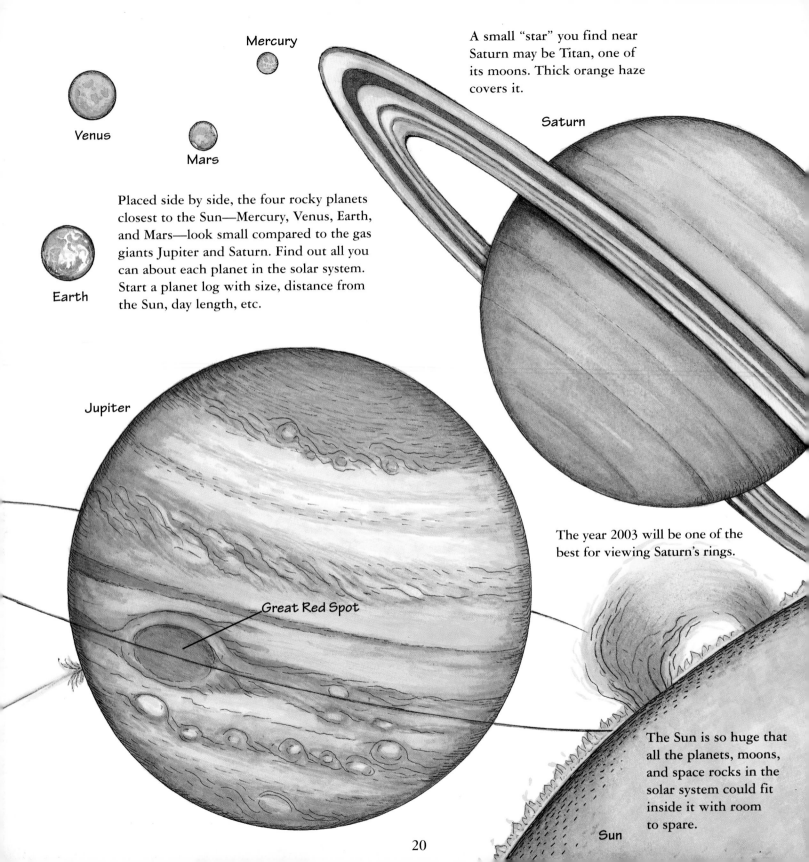

Mercury

Venus

Mars

A small "star" you find near Saturn may be Titan, one of its moons. Thick orange haze covers it.

Saturn

Placed side by side, the four rocky planets closest to the Sun—Mercury, Venus, Earth, and Mars—look small compared to the gas giants Jupiter and Saturn. Find out all you can about each planet in the solar system. Start a planet log with size, distance from the Sun, day length, etc.

Earth

Jupiter

The year 2003 will be one of the best for viewing Saturn's rings.

Great Red Spot

The Sun is so huge that all the planets, moons, and space rocks in the solar system could fit inside it with room to spare.

Sun

20

Sojourner

All eyes were on Mars during the summer of 1997. There the robot rover *Sojourner* rolled down a ramp and roamed among the red rocks.

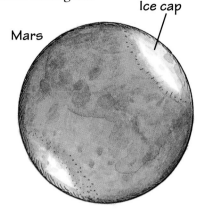
Mars
Ice cap

The Wanderers

Suppose one night a new star appears in your square, shining more brightly than any star around it. What should you do? Call a news station and report a supernova? Not so fast. First do a little detective work.

Look carefully at the star with your eyes, then with binoculars or a telescope. Note down every detail: size, shape, brightness, color, and anything unusual about it. For instance, does it twinkle? Get out your notebook and add the new star to your small square, in its exact spot. By now, you know the constellations in your square. Is the new star in or near any of these: Taurus, Gemini, Cancer, Leo, Virgo, Libra, Scorpius, Sagittarius, Capricornus, Aquarius, Pisces, Aries? If so, prepare yourself. Most likely your "exploding star" is a planet.

Including Earth, nine planets circle the Sun. You can spot five with your eyes: Mercury, Venus, Mars, Jupiter, and Saturn. The other three—Neptune, Uranus, and Pluto—are so far away that you'll need to check a newspaper or sky magazine to find out when they can be seen. Then you'll need a telescope to locate them.

Planets don't shine as stars do. You can see a planet

For three months every two years, Mars is closest to Earth. Wait for it. It's the best chance to see the surface, unless a Martian dust storm spoils the view.

The planets and the Moon always appear on or near the dotted line in the small square. This line is called the ecliptic.

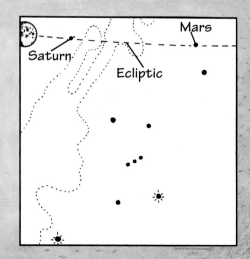
Mars
Saturn
Ecliptic

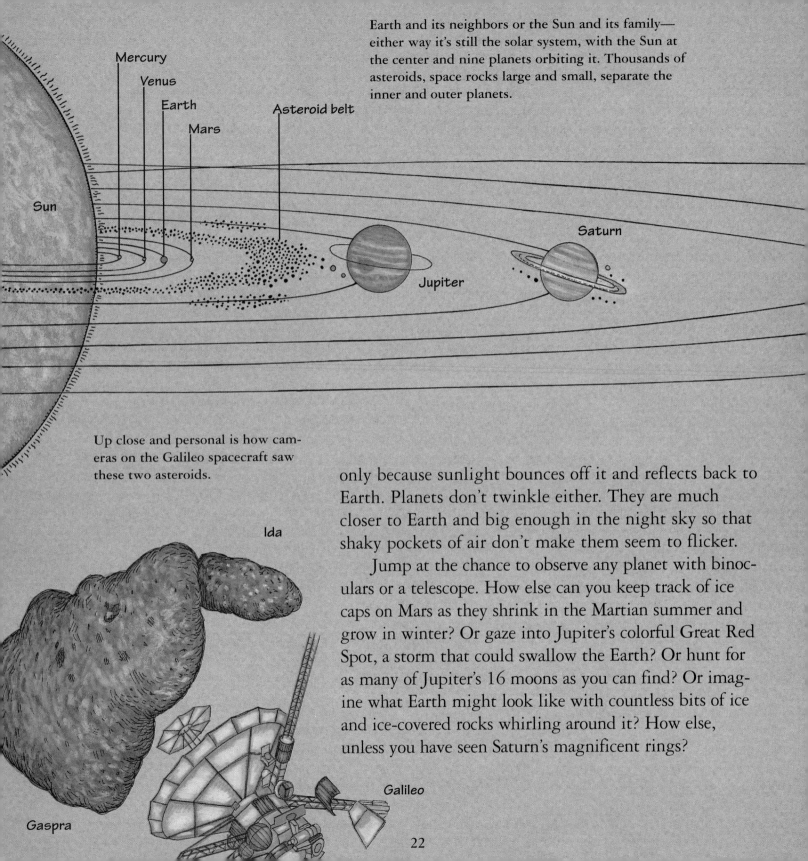

Earth and its neighbors or the Sun and its family—either way it's still the solar system, with the Sun at the center and nine planets orbiting it. Thousands of asteroids, space rocks large and small, separate the inner and outer planets.

Sun

Mercury

Venus

Earth

Mars

Asteroid belt

Jupiter

Saturn

Up close and personal is how cameras on the Galileo spacecraft saw these two asteroids.

Ida

Gaspra

Galileo

only because sunlight bounces off it and reflects back to Earth. Planets don't twinkle either. They are much closer to Earth and big enough in the night sky so that shaky pockets of air don't make them seem to flicker.

Jump at the chance to observe any planet with binoculars or a telescope. How else can you keep track of ice caps on Mars as they shrink in the Martian summer and grow in winter? Or gaze into Jupiter's colorful Great Red Spot, a storm that could swallow the Earth? Or hunt for as many of Jupiter's 16 moons as you can find? Or imagine what Earth might look like with countless bits of ice and ice-covered rocks whirling around it? How else, unless you have seen Saturn's magnificent rings?

22

Because Mercury and Venus are closer to the Sun than Earth is, you'll see them only in the western sky after sunset or the eastern sky before sunrise. Venus reflects so much sunlight that it has been called the evening or morning star.

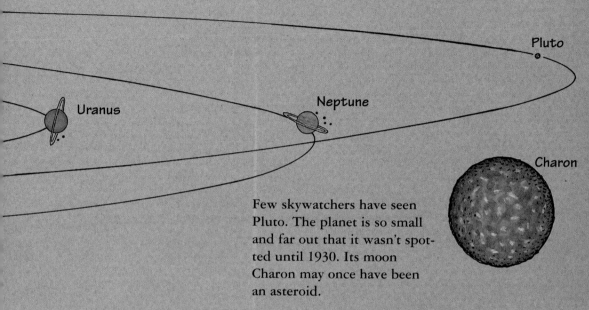

Uranus

Neptune

Pluto

Charon

Few skywatchers have seen Pluto. The planet is so small and far out that it wasn't spotted until 1930. Its moon Charon may once have been an asteroid.

When and how you can see different planets depends on where they are in the orbits around the Sun, compared with Earth's. If Jupiter, for example, is on the other side of the Sun from Earth, you won't be able to see it. Likewise, there will be times when Saturn's rings seem to disappear. That's a clue that you are viewing them edge-on. Not to worry. They will return to their full glory as Saturn moves along its path around the Sun.

From the night a planet shows up in your small square, keep tabs on its position. It may start out next to a certain star. A month later it may be in a nearby constellation. Still later, it will move again, while the pattern of stars always remains the same. Such a night-sky wanderer is a dead giveaway that you have pinpointed a planet. In fact, "planet" comes from the Greek word for wanderer.

Loop the Loop

Mars is farther from the Sun than Earth is. It takes the Red Planet almost twice as long to orbit the Sun than Earth does. If reddish-looking Mars enters your square, watch it over a few months. Every time you see it, draw in your notebook its location among the stars. Mark down the date, too, so you can connect the points in the proper order. Do you find that Mars moves in one direction most of the time? That's east. Then does it turn and loop back west? And then loop again back east? Truth is, Mars never shifts into reverse. It always wanders east across the night sky.

To solve the mystery of the loops, select a tree in your yard or in the park to be the Sun. You be the Earth and circle the tree about 5 feet (1.5 m) from it.

Ask a friend to be Mars and to circle about 8 feet (2.4 m) from the tree but walk half as fast as you do. As you both keep circling, look at your friend. When does your friend seem to slow down, go backward, or speed up?

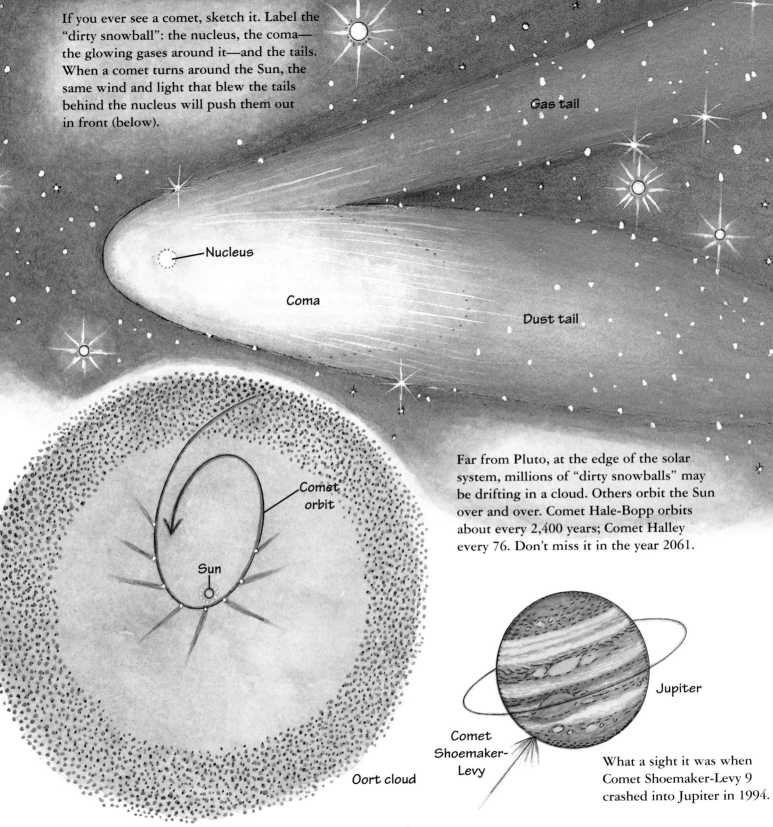

If you ever see a comet, sketch it. Label the "dirty snowball": the nucleus, the coma—the glowing gases around it—and the tails. When a comet turns around the Sun, the same wind and light that blew the tails behind the nucleus will push them out in front (below).

Gas tail

Nucleus

Coma

Dust tail

Comet orbit

Sun

Oort cloud

Far from Pluto, at the edge of the solar system, millions of "dirty snowballs" may be drifting in a cloud. Others orbit the Sun over and over. Comet Hale-Bopp orbits about every 2,400 years; Comet Halley every 76. Don't miss it in the year 2061.

Jupiter

Comet Shoemaker-Levy

What a sight it was when Comet Shoemaker-Levy 9 crashed into Jupiter in 1994.

24

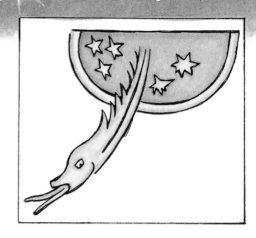

Comet Hale-Bopp

Name That Comet

Have you heard of Halley? How about Hale-Bopp? Sound like names of people? They are. But they are also the names of comets. And when there's a comet lighting up some small square of the night sky, it causes a sensation around the world.

If you were lucky enough to see Comet Hale-Bopp a few years ago, you already know why. There it was, night after night, a fuzzy, glowing ball with a long, streaming tail. Where did it come from? Far beyond Pluto. Where was it going? Around the Sun and back out into space again.

A comet starts out on its journey as a dark chunk of ice and dust a few miles (km) wide. Think of it as a super–hard-packed, dirty snowball. If the "dirty snowball" falls into the solar system, it can go into an egg-shaped orbit around the Sun.

By the time a comet swings by Earth, the Sun's heat is already changing some of the ice into gases and freeing some of the dust. These create the fuzzy, glowing ball. At the same time, the Sun's light and a wind of particles rushing out of the Sun blow the glowing gases and dust into two long tails. How long? Often millions and millions of miles (km)!

Every year, skywatchers discover about a dozen new comets. Most glow only faintly. If you are the first to see and report a new comet, that comet will bear your name.

Comets as evil omens? Hard to believe. Yet for centuries, people blamed comets for wars, floods, disease, and famine. The Aztecs, in Mexico, may have drawn a comet (above) when they were conquered by Spain. To the ancient Chinese, different comets (left) brought different ills. Comet Halley showed up in 1066, when the Normans invaded England. Can you find this famous comet below?

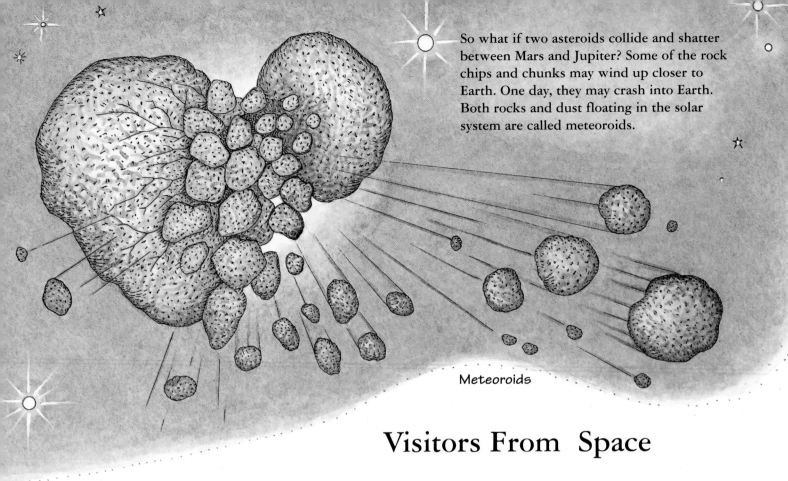

So what if two asteroids collide and shatter between Mars and Jupiter? Some of the rock chips and chunks may wind up closer to Earth. One day, they may crash into Earth. Both rocks and dust floating in the solar system are called meteoroids.

Meteoroids

Visitors From Space

Earth is on a collision course, but no one cares. After all, it's only a bunch of space rocks that keep getting in Earth's way. Sure, some are boulder size. But most are smaller than jellybeans. Earth versus rocks: what a joke! Or is it?

One night you may have a ringside seat as Earth faces off against a rocky visitor from space in your small square. True, you won't witness the moment of impact as a rock smacks into Earth's blanket of air. If the rock bounces back into space, then it's a knockout. But if the rock keeps going, it will rub against the air at high speed. If the rock is small, it can spell doom. The rubbing will heat the rock so much that it burns. Before your very eyes, the fiery rock will shoot across the night sky as a streak of light called a meteor or a "shooting star" and disappear.

Shouldn't Meteor Crater in Arizona be "Meteorite"? That's what crash-landed there over 25,000 years ago, exploded, and left this gaping hole.

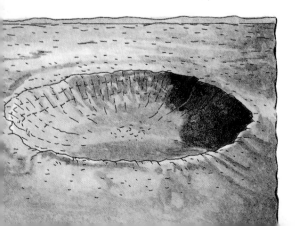

If the rock is large, it may not burn up. Instead the fireball may keep going and crash on the land or sea. Then it is called a meteorite. About 65 million years ago a giant meteorite (maybe even an asteroid) slammed into Earth. It exploded with the force of many megaton nuclear bombs and threw enough dust into the air to block the Sun for months. It was no joke, for it may have brought about the end of the dinosaurs and many other kinds of life. With luck, no other deadly space rock will visit for a long, long time.

You cannot plan ahead to see a shooting star. One may happen in your square at any time. But meteor showers are a different story. They take place at certain times every year when Earth crosses a comet's path. What a show, as a dozen or more burning dust specks the comet left behind zoom across the night sky hour after hour! No wonder skywatchers welcome these visitors with open eyes.

If it's a meteor shower in Orion, the date must be October 20 or 21. That's about when a shower takes place there every year, as Earth crosses the path of Comet Halley.

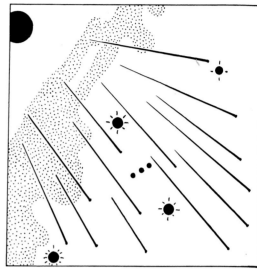

When this space rock crashes, a thick curtain of dust will shoot up and blacken the sky for many months. Without sunlight, Earth will cool and plants will die. Then plant eaters. Then meat eaters. It will be curtains for dinosaurs forever, or so many people think.

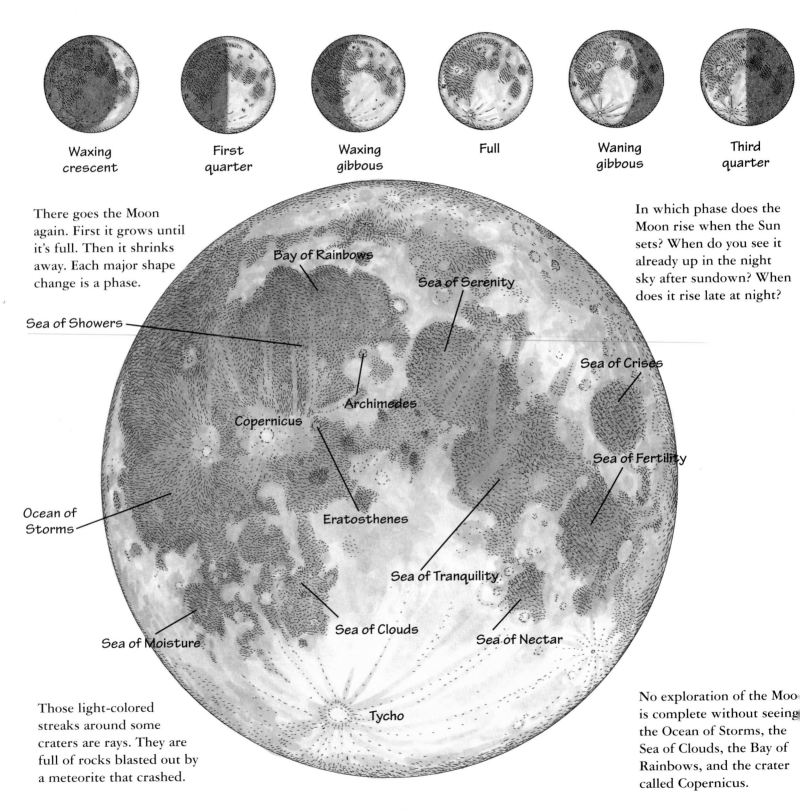

Waxing crescent

First quarter

Waxing gibbous

Full

Waning gibbous

Third quarter

There goes the Moon again. First it grows until it's full. Then it shrinks away. Each major shape change is a phase.

Sea of Showers

Bay of Rainbows

Sea of Serenity

In which phase does the Moon rise when the Sun sets? When do you see it already up in the night sky after sundown? When does it rise late at night?

Sea of Crises

Archimedes

Copernicus

Sea of Fertility

Ocean of Storms

Eratosthenes

Sea of Tranquility

Sea of Moisture

Sea of Clouds

Sea of Nectar

Tycho

Those light-colored streaks around some craters are rays. They are full of rocks blasted out by a meteorite that crashed.

No exploration of the Moo is complete without seeing the Ocean of Storms, the Sea of Clouds, the Bay of Rainbows, and the crater called Copernicus.

28

Waning crescent

Once more, the action is along the dotted line. That's where you will find the Moon in the small square. If the Moon is in your square, so, too, is a stretch of the ecliptic.

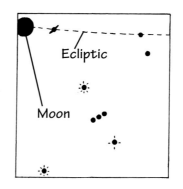

Ecliptic

Moon

Moonwatch

No matter how you look at it, the Moon is special. At a mere 240,000 miles (384,000 km) or so away, it is Earth's nearest neighbor in space. Nothing else in the night sky is so big, so bright, and so easy to see.

If you are in the mood for mountains, check out the Moon. Curious about craters? The Moon is loaded with them, and you can stare deep inside their towering walls. How about a bit of space history? Let your binoculars whisk you up where astronauts landed and took those famous steps on lunar soil. So go ahead. Give in. Even if it's not in your small square, take a peek at the Moon whenever you can.

Where in the sky you'll find the Moon keeps changing. So does the Moon's shape. These happen because the Moon circles the Earth about once every month.

As the Earth turns, the Moon, like the stars, seems to rise in the east and set in the west. If the Moon gave off its own light, it would always look the same. But it doesn't. Moonlight is sunlight that hits the Moon and bounces back to Earth. The Sun shines on only half the Moon at once. Where it does, it is Moon day. Where it does not, it is Moon night.

Terminator

For a really spectacular view of the Moon's mountains and craters, focus your binoculars on the terminator. It may pass through the Sea of Tranquility, site of the first moonwalk.

Phase In

Place an orange or a ball on a table as a stand-in for the Moon. Ask a friend to be the Sun and shine a flashlight on it. Darken the room.

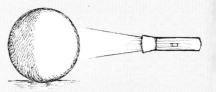

Observe the ball first from behind your friend, then as you slowly walk around it. How does the lighted part of the ball that you can see change as you circle around? Can you see why the Moon seems to change as it goes through its phases?

One Good Turn

Earth turns around once every 24 hours. The Moon rotates, too—every 27.3 days. That's also how long the Moon takes to circle the Earth once. What does this mean to you? Stand in the middle of a room. Have a friend circle slowly around you and keep turning so that he or she faces all four of the room's walls by the time the circle is complete. Do you ever see your friend's back during the circling? For the same reason you can see only one side of the Moon as it circles the Earth.

29

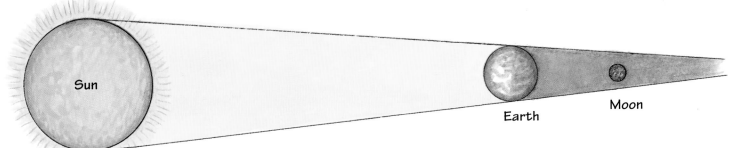

Sun

Earth

Moon

If you see a full Moon start to disappear, a lunar eclipse is taking place. Earth has lined up exactly between the Moon and the Sun, blocking the sunlight.

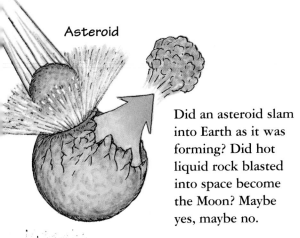

Asteroid

Did an asteroid slam into Earth as it was forming? Did hot liquid rock blasted into space become the Moon? Maybe yes, maybe no.

It's a solar eclipse when the Moon blocks sunlight from Earth so that somewhere on Earth, day turns to night. NEVER look at the blackened Sun in an eclipse. It still can damage your eyes.

Most of the time that you look at the Moon, you can see where Moon night ends and day begins. This dividing line is called the terminator. As the Moon moves around the Earth, the terminator doesn't stay in one place. Follow the terminator and the Moon's ever-changing shapes, or phases, will unfold. Just keep in mind what's really changing is how much of the sun-lit half of the Moon you can see each night from Earth. When the Moon is opposite the Sun, the half you see is fully lit. When the Moon is in front of the Sun, the lighted half faces away from you. Though you see no Moon, this phase is called new Moon.

Whenever you Moonwatch, take note of everything. Where is the Moon in the sky? Where is the terminator? Which phase do you find best for viewing mountains and craters? What stands out most when the Moon is full? Do the darkish areas look like water to you? They did to the Moon mapper who named them *maria,* Latin for "seas." Indeed, the *maria* were "seas" once—lava seas that hardened into rock 3.5 billion years ago. Whatever you do—sketch it, map it, explore it—the Moon is yours for the watching.

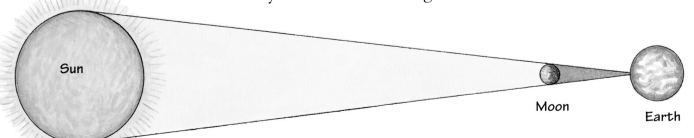

Sun

Moon

Earth

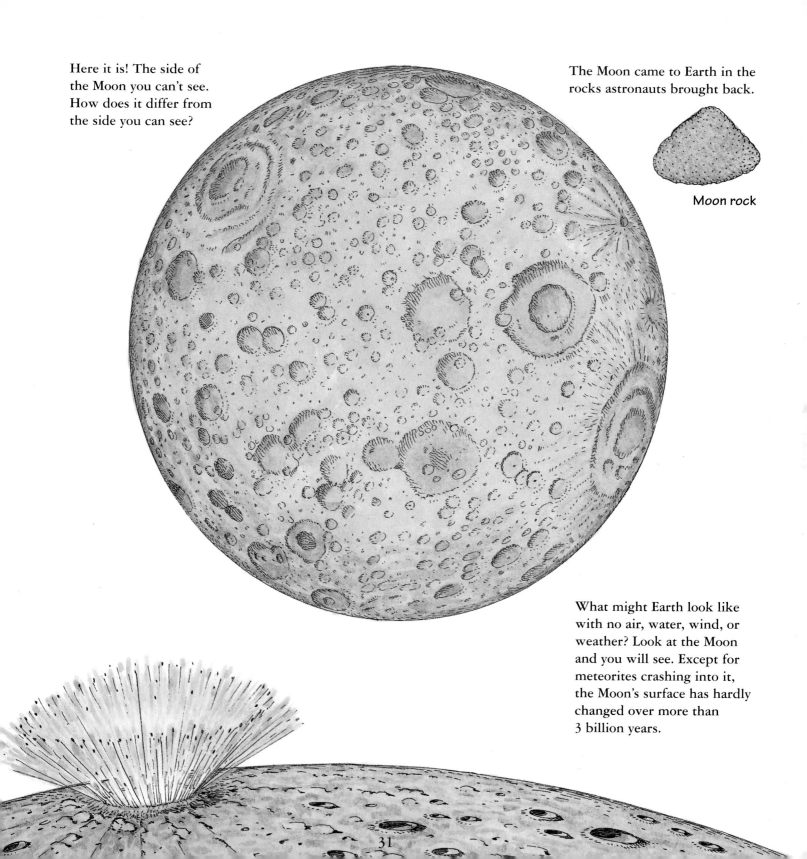

Here it is! The side of the Moon you can't see. How does it differ from the side you can see?

The Moon came to Earth in the rocks astronauts brought back.

Moon rock

What might Earth look like with no air, water, wind, or weather? Look at the Moon and you will see. Except for meteorites crashing into it, the Moon's surface has hardly changed over more than 3 billion years.

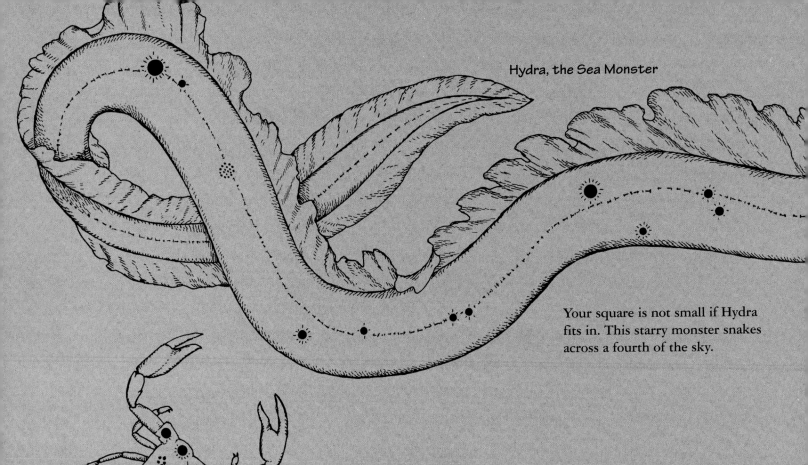

Hydra, the Sea Monster

Your square is not small if Hydra fits in. This starry monster snakes across a fourth of the sky.

Antares

Scorpius, the Scorpion

Are any stars in these constellations part of your small square? If not, do any take the place of stars that were once there but are no longer? As your square changes with the seasons, how do new stars compare with the old?

The Ride of Your Life

You are taking the ride of your life and you may not even know it. In the last hour alone, Earth has moved you about 65,000 miles (105,000 km) along its path around the Sun. During that same hour, the Sun pulled you, the Earth, and the rest of the solar system with it about 500,000 miles (800,000 km) as it speeds around the center of the Milky Way Galaxy once every 200 million years. Meanwhile, the entire Milky Way, including you, has traveled millions more miles (km) as it circles other galaxies. Get the picture? There's a whole lot of moving going on. As you sit in one place and read this, you can't feel any of that motion. But you can see some proof of it by exploring your small square night after night, month after month.

32

Altair

Aquila,
the Eagle

Pegasus,
the Winged Horse

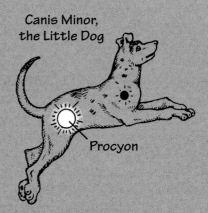

Canis Minor,
the Little Dog

Procyon

With just two stars,
the Little Dog is little.
But it's not hard to
find, with its one
bright star.

When you skywatch, do you go out at the same time each night? Do you hold your small square viewer in exactly the same spot? If so, you have already discovered that the stars you saw in the viewer the first night moved out of the square weeks ago. Other stars in other constellations replaced them. Then they, too, were replaced.

Or did you decide to stick with the stars you first saw in your square? If that answer is yes, and you have been sky-watching around the same time each night, then every time you have gone out, you have been moving your viewer a little to find those same stars. In fact, you may have followed them clear across the sky until they no longer rise while you are skywatching.

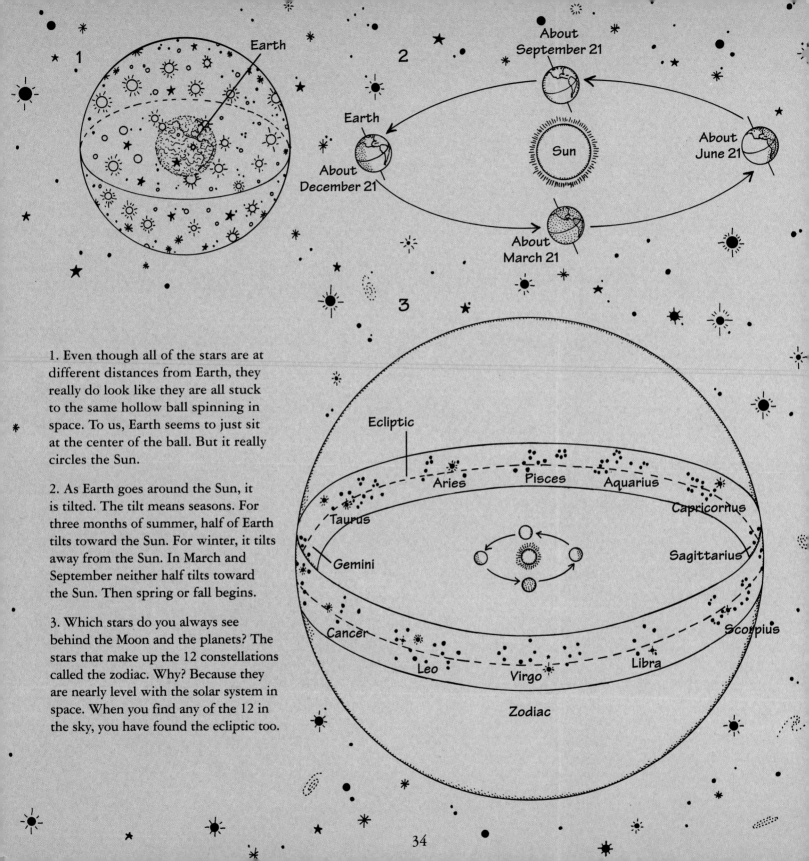

1 Earth

2 About September 21 · Earth · Sun · About June 21 · About December 21 · About March 21

3 Ecliptic · Aries · Pisces · Aquarius · Taurus · Capricornus · Gemini · Sagittarius · Cancer · Scorpius · Leo · Virgo · Libra · Zodiac

1. Even though all of the stars are at different distances from Earth, they really do look like they are all stuck to the same hollow ball spinning in space. To us, Earth seems to just sit at the center of the ball. But it really circles the Sun.

2. As Earth goes around the Sun, it is tilted. The tilt means seasons. For three months of summer, half of Earth tilts toward the Sun. For winter, it tilts away from the Sun. In March and September neither half tilts toward the Sun. Then spring or fall begins.

3. Which stars do you always see behind the Moon and the planets? The stars that make up the 12 constellations called the zodiac. Why? Because they are nearly level with the solar system in space. When you find any of the 12 in the sky, you have found the ecliptic too.

Either way, one thing is for sure: As you travel aboard planet Earth in its orbit around the Sun, the night sky is ever-changing. That's because the ride Earth takes you on lets you look out at different parts of the Milky Way Galaxy—at least until the ride begins again at year's end.

Think of the ride as a merry-go-round. You are on one of the horses. Call it Earth. At the center of the ride is a very big, very bright light. Call it Sun. The light's blinding glare—call it Day—stops you from seeing anything behind it. But you can look straight out in the opposite direction. Call it Night. There you can see lots of different houses and trees. Call them Stars.

The ride begins. As you circle once around the center and look out the other way, the view always changes. When you complete one circle—call it Year—and begin another, the view repeats too.

Enjoy the view as Earth takes you on the ride of your life. And be glad that it repeats year after year. There's so much to see, so many stars to get to know, and so many surprises, such as shooting stars and soaring comets.

The more you explore the night sky, the less you will need to rely on your small square viewer to help you locate stars. Constellations will become old friends, easy to pick out and great to have along for the ride.

North, You Say?

Do you live north of the Equator? If so, how far? You could look on a map, but why not let the North Star help you? First, find Polaris in the night sky. (See page 4 if you forgot how.) Then hold both your arms straight out in front of you as shown. Raise one arm

until it points to Polaris. If that arm points directly above your head, you live near the North Pole. If your arm hardly rises, you live near the Equator. The closer the arm is to straight up, the farther north you live.

Need a Telescope?

Nothing pleases skywatchers more than showing off the wonders of the night sky. They often form clubs so members can bring their telescopes and stargaze together. You can find out if there is a club near you by looking in *Sky & Telescope* or *Astronomy* magazines. If there's a planetarium in your city, it may have a list. When you join such a club, you'll meet people who can teach you how to use a telescope and may let you look through theirs.

Moon

Saturn

Mars

Meteor

Rosette Nebula (Monoceros)

Cone Nebula (Monoceros)

Nebula NGC 2174 (Orion)

Name That Constellation

Remove the lid from a shoe-box. Punch a hole in one end of the box, as shown. Then cut out the opposite end.

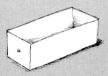

Replace the lid on top of the box. Cut a slit in the lid close to the end that is now open. Measure the length of the slit

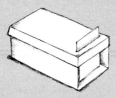

as well as the height of the box. Measure and cut rectangles out of cardboard or heavy black paper so each is as long as the slit and an inch (2.5 cm) higher than the box height.

Using the sky charts in this book, draw the star patterns of different constellations, one on a rectangle. Make a hole in each drawn star with a pin. Slide the rectangle into the slit and hold the box up to a light. When you look through the hole, can you see the "constellation"? If not, enlarge the pinholes until you can. See if your friends can name each constellation you drew.

The Sky's the Limit

The universe is big. It's hot. Some days it's everywhere you turn. It makes headlines when planets are spotted orbiting distant stars and when millions of people visit a website to get a glimpse of spectacular photos being beamed back from Mars. The universe is creating a stir as skywatchers, with the help of the Hubble Space Telescope, get a better look at Pluto and at newly discovered galaxies that seem to go on without end.

Once you see what's out there, you'll want to see more, longer, and more clearly. You'll want to find out where it all came from, where it's all going, and how it all works. Your list of unanswered questions will grow and grow: Will another asteroid ever crash into Earth? When will Betelgeuse self-destruct? If there's life on Earth, why not somewhere else in the universe? Just remember: At one time, skywatchers asked what makes stars rise and set and why the Moon changes shape. Now you know.

No one can tell what the future will bring. Not even someone who studies the stars. But there is a chance that many of your questions will be answered during your lifetime. Until then, explore one small square. Explore as many as you can. The sky's the limit.

One Small Square of Night Sky

Figure out the constellation in which you would see the Moon on this night. Saturn. Mars. The meteor.

36

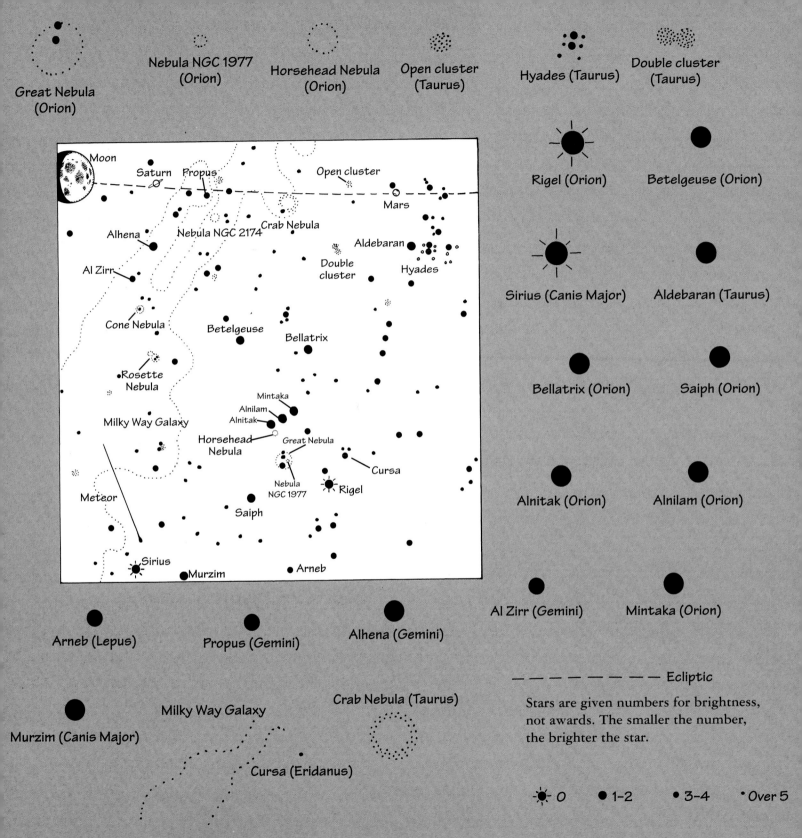

Great Nebula (Orion)

Nebula NGC 1977 (Orion)

Horsehead Nebula (Orion)

Open cluster (Taurus)

Hyades (Taurus)

Double cluster (Taurus)

Rigel (Orion)

Betelgeuse (Orion)

Sirius (Canis Major)

Aldebaran (Taurus)

Bellatrix (Orion)

Saiph (Orion)

Alnitak (Orion)

Alnilam (Orion)

Al Zirr (Gemini)

Mintaka (Orion)

Arneb (Lepus)

Propus (Gemini)

Alhena (Gemini)

Murzim (Canis Major)

Milky Way Galaxy

Crab Nebula (Taurus)

Cursa (Eridanus)

— — — — — — — — — — — Ecliptic

Stars are given numbers for brightness, not awards. The smaller the number, the brighter the star.

-☀- 0   ● 1-2   ● 3-4   • Over 5

Map labels: Moon, Saturn, Propus, Open cluster, Mars, Alhena, Nebula NGC 2174, Crab Nebula, Aldebaran, Al Zirr, Double cluster, Hyades, Cone Nebula, Betelgeuse, Bellatrix, Rosette Nebula, Mintaka, Alnilam, Alnitak, Milky Way Galaxy, Horsehead Nebula, Great Nebula, Cursa, Nebula NGC 1977, Rigel, Meteor, Saiph, Sirius, Murzim, Arneb

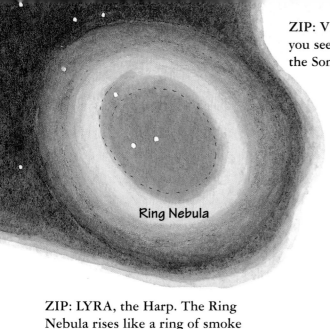

Ring Nebula

ZIP: VIRGO, the Virgin. Can you see why this galaxy is named the Sombrero?

Sombrero Galaxy

Don't Forget the "Zip Code"

Everything in the night sky has an address that can help you locate it. The address may be a name, a number, or letters and numbers. Whatever it is, it always ends with the name of a constellation. That's your tip-off where to start looking in the night sky. It's like a Zip Code in space.

Use the sky charts on the following pages to identify constellations as the seasons change. Consult a star atlas or a field guide to the night sky for the names of stars, nebulas, and galaxies. You'll find one at the library. As you list the names of everything you've seen, don't forget the "zip."

ZIP: LYRA, the Harp. The Ring Nebula rises like a ring of smoke from a dying star.

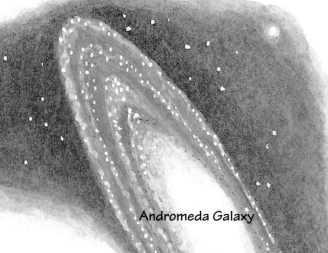

Andromeda Galaxy

ZIP: ANDROMEDA, the Princess. What's 13 quintillion miles (21 quintillion km) away? A spiral galaxy you can see on a dark night even from a city. That's 13,000,000,000,000,000,000!

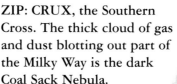

Pleiades

ZIP: TAURUS, the Bull. Of the 500 stars in this open cluster, how many can you spot with just your eyes?

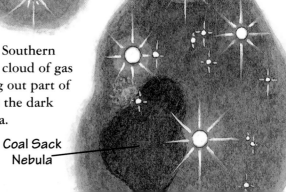

ZIP: CRUX, the Southern Cross. The thick cloud of gas and dust blotting out part of the Milky Way is the dark Coal Sack Nebula.

Coal Sack Nebula

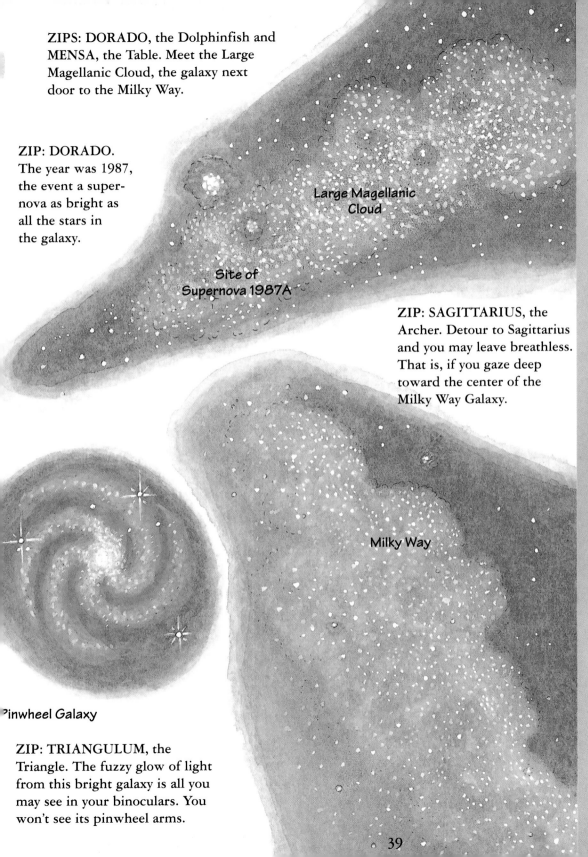

ZIPS: DORADO, the Dolphinfish and MENSA, the Table. Meet the Large Magellanic Cloud, the galaxy next door to the Milky Way.

ZIP: DORADO.
The year was 1987, the event a super-nova as bright as all the stars in the galaxy.

Large Magellanic Cloud

Site of
Supernova 1987A

ZIP: SAGITTARIUS, the Archer. Detour to Sagittarius and you may leave breathless. That is, if you gaze deep toward the center of the Milky Way Galaxy.

Milky Way

Pinwheel Galaxy

ZIP: TRIANGULUM, the Triangle. The fuzzy glow of light from this bright galaxy is all you may see in your binoculars. You won't see its pinwheel arms.

Which Constellation Is It?

The skycharts on the following pages will help you iden-tify most of the constellations in the night sky. Select either the northern hemisphere chart or the southern hemisphere chart, depending on where you live.

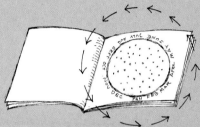

To use a chart, you need to turn the book so that the current month is closest to you. Try to match the star patterns you see in the sky to those above the name of the month. The brightest stars and the red-orange stars will help you locate many constellations. So will the Milky Way, if you can see it. Be sure to read the hints on the following pages for using each chart. And remember: The sky keeps changing as the hours and the seasons pass.

School is out and nights are warm. Three bright stars form the Summer Triangle in the sky. One is Deneb, in Cygnus. The second is Altair, in Aquila. And the third is Vega, in Lyra. Antares, a bright-red star, will lead you to Scorpius, the deadly scorpion.

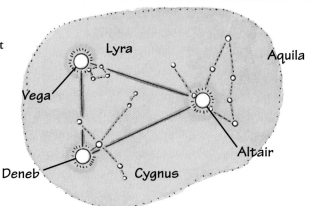

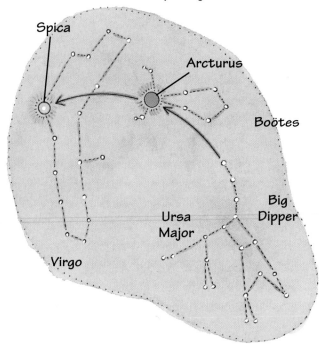

Searching for the Big Dipper? It's part of the constellation Ursa Major, the Big Bear. If you follow the curve of the Big Dipper's handle, you can "arc" to Arcturus, the brightest star in the Boötes. Keep the curve going and you will "speed" to Spica, the bright star in Virgo.

Northern Hemisphere

This chart shows the stars you can see if you live in the northern hemisphere. Whenever you skywatch, first find Polaris, the North Star. It's in the north part of your sky. Face Polaris and the south part of your sky will be at your back.

Turn the book so that the current month is closest to you. No matter how you turn the chart, Polaris is always at the center. The stars around Polaris on the chart are the ones you will find in the northern part of your sky. The stars above the name of the month are the ones you will find in the southern part of your sky.

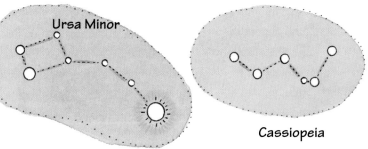

Shapes are the key to the sky in autumn. There's the square in Pegasus, the W-shape of Cassiopeia, and the Little Dipper that is Ursa Minor.

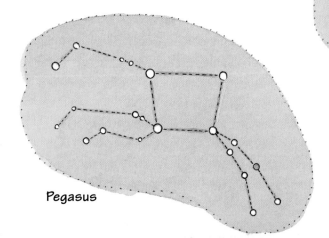

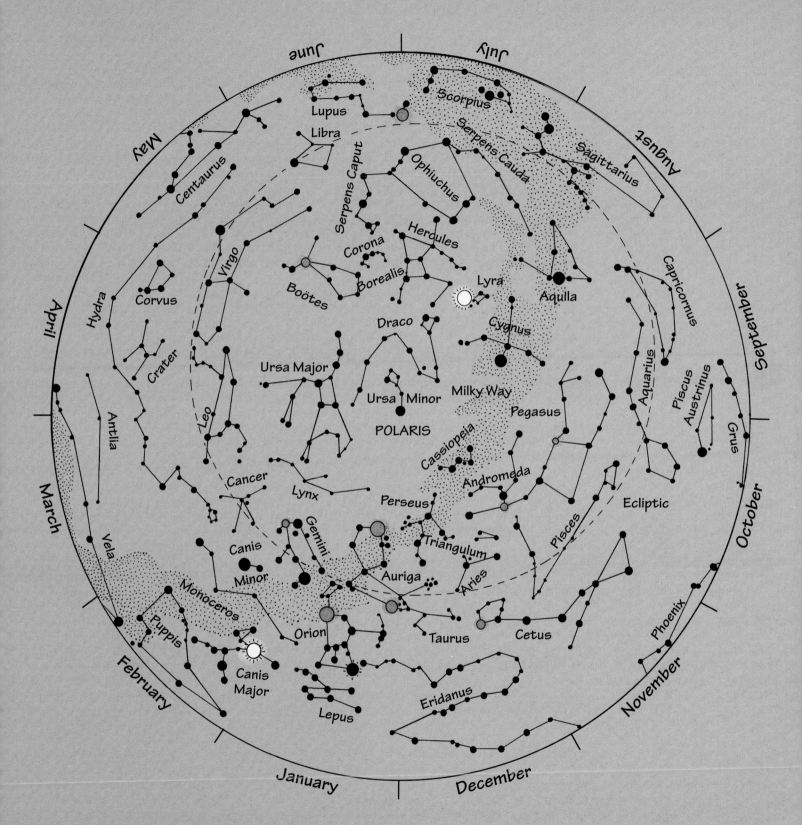

41

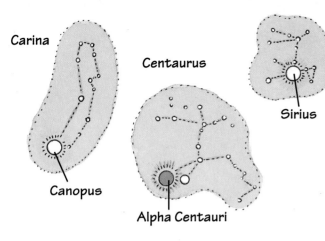

Carina

Centaurus

Canis Major

Canopus

Sirius

Alpha Centauri

Southern Hemisphere

No, Orion's not standing on his head. Nor are you if you are seeing it from the southern hemisphere. That's how it looks to the people living there.

If you skywatch from the southern hemisphere, turn this chart so that the current month is closest to you. The stars at the center of the chart are the ones you will find in the southern part of your sky. The stars above the name of the month are the ones you will find in the northern part of your sky. These include Orion and all the stars around it.

You're in luck. Not only Sirius, the brightest star, is awaiting you in the night sky. So are the second-brightest—Canopus, in the constellation Carina—and the third-brightest—Alpha Centauri, in Centaurus.

So you can't see the North Star from the southern hemisphere. But you still can "arc" to Arcturus and "speed" to Spica by following the curve of the Big Dipper's handle.

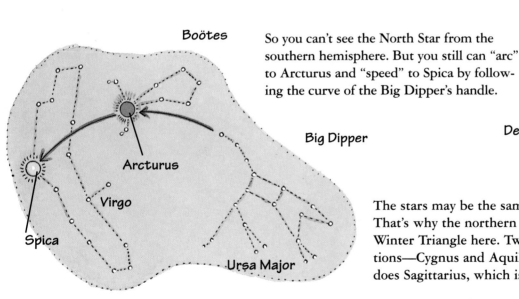

Boötes

Arcturus

Virgo

Spica

Big Dipper

Ursa Major

Vega

Lyra

Aquila

Deneb

Cygnus

Altair

The stars may be the same, but the seasons aren't. That's why the northern Summer Triangle is the Winter Triangle here. Two of the triangle constellations—Cygnus and Aquila—overlap the Milky Way. So does Sagittarius, which is rich in nebulas and clusters.

Match the square part of Pegasus to the real square in the sky and you'll have an easier time finding other nearby constellations. This is especially true for Andromeda, since one corner star of the square belongs in it, not in Pegasus.

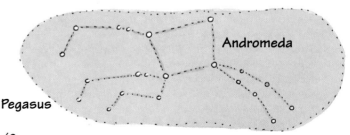

Andromeda

Pegasus

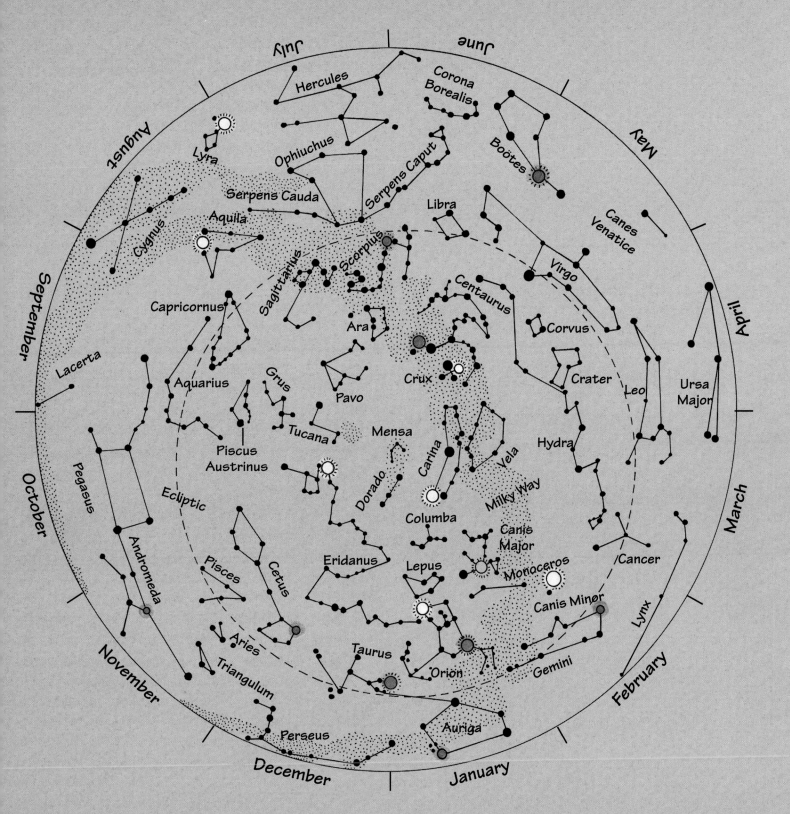

43

Index

Orion's belt

Light

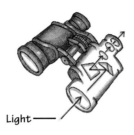

Index

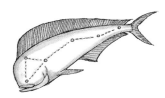

Index

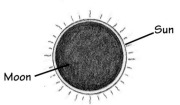

Index

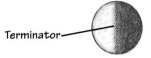

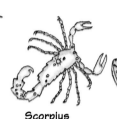

Aquarius
(The Water Carrier)

Capricornus
(The Sea Goat)

Sagittarius
(The Archer)

Scorpius
(The Scorpion)

Libra
(The Scales)

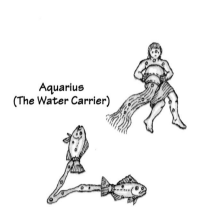

Pisces
(The Fishes)

ZODIAC

Virgo
(The Virgin)

Aries
(The Ram)

Taurus
(The Bull)

Gemini
(The Twins)

Cancer
(The Crab)

Leo
(The Lion)

Further Reading

Look for the following in a library or bookstore:

Astronomy magazine

Sky & Telescope magazine

Weekly skycharts in newspapers

365 Starry Nights by C. Raymo, Prentice-Hall, Englewood Cliffs, NJ, 1982.

Stars, Clusters, and Galaxies by J. Gustafson, Julian Messner, New York, NY, 1992.

A Field Guide to the Stars and Planets by D. H. Menzel and J. M. Pasachoff, Houghton Mifflin, Boston, MA, 1983.